NATURE WALK

The Squirrel

NATURE WALK

NATURE WALK

The Squirrel

James V. Bradley

CHELSEA CLUBHOUSE

An Imprint of Chelsea House Publishers

THE SQUIRREL
© 2006 by Infobase Publishing

Chelsea Clubhouse
An imprint of Infobase Publishing
132 West 31st Street
New York NY 10001

Library of Congress Cataloging-in-Publication Data

Bradley, James V. (James Vincent), 1931–
  The squirrel / James V. Bradley.
    p. cm. — (Nature walk)
  Includes bibliographical references and index.
  ISBN 0-7910-9116-3 (hardcover)
  1. Squirrels—Juvenile literature. I. Title. II. Series: Bradley,
James V. (James Vincent), 1931– Nature walk.
  QL737.R68B73 2006
  599.36—dc22   2006011764

Chelsea House books are available at special discounts when purchased in bulk quantities for businesses, associations, institutions, or sales promotions. Please call our Special Sales Department in New York at (212) 967-8800 or (800) 322-8755.

You can find Chelsea House on the World Wide Web at
http://www.chelseahouse.com

TEXT AND COVER DESIGN by Takeshi Takahashi
ILLUSTRATIONS by William Bradley
SERIES EDITOR Tara Koellhoffer

Printed in the United States of America

_PKG 10 9 8 7 6 5 4 3 2 1

FEB 0 3 2007

This book is printed on acid-free paper.

All links and Web addresses were checked and verified to be correct at the time of publication. Because of the dynamic nature of the Web, some addresses and links may have changed since publication and may no longer be valid.

TABLE OF CONTENTS

Chapter 1

Introduction to Squirrels

Origin of Squirrels

SQUIRRELS ARE **rodents,** along with rats, mice, ground-hogs, prairie dogs, porcupines, muskrats, hamsters, beavers, and chipmunks. The oldest fossil rodents are about 50 million years old, and the oldest fossil squirrels in the Americas date from about 28 to 30 million years ago (during the late Eocene Epoch). These fossilized skeletons, especially the skulls, are strikingly similar to those of the squirrels that are alive today. Evidently, after **adapting** to life in trees, the squirrel's skeleton has changed little over time. We know that these early tree squirrels shared the forest with small three-toed horses and other mammals, most of which eventually became **extinct**.

Think of what it must have been like for the first rodents to explore life in trees. They found a new habitat that was wide open with new sources of food and—best of all—no **predators**. Somewhat similar situations sometimes occur today when new **organisms** (such as the English sparrow or the Asian long-

Squirrels adapted over millions of years to be able to live in trees.

horned beetle), are introduced into the United States. With new sources of food and shelter and a lack of predators, they multiply and spread, and often become the strongest members of the community. This is why there are strict laws about which kinds of living things may be brought into the country.

Where Squirrels Live Today

Tree squirrels are found in North and South America, Europe, Asia, Africa, Japan, Indonesia, and New Zealand, but not Australia or Antarctica. The largest squirrel is the giant squirrel of India, which has a body length of 3 feet (0.9 m). The smallest is the pygmy squirrel of Africa, which is only about 5 inches (12.7 cm) long.

Adaptation to Life in the Trees

Even the oldest skulls of fossil tree squirrels closely resemble the skulls of modern squirrels. The basic plan of the squirrel skeleton was a simple one and did not change much with time. The brain of the squirrel is about the size of a walnut. In fact, it even looks like a walnut.

Because **gnawing** is essential for rodents, their **incisors** can grow all the time. Squirrels' incisors may grow up to 6 inches (15 cm) in one year. If not kept worn down by frequent gnawing, they would overlap and the squirrel would not be able to chew its

The skull of the modern squirrel has not changed much from the skull of its relatives millions of years ago.

food. This may happen, for example, when a squirrel injures its jaw. Only the front of an incisor is covered by hard tooth enamel. The back is made of softer **dentine**. The back of the tooth wears away faster than the front. This causes the tooth to take on a chisel-like shape that is ideal for gnawing on woody plant tissue.

The lower incisors, which are about twice as long as the upper incisors, have been modified to make it easier for squirrels to crack nuts. Muscles in the lower jaw contract to spread the two lower incisors apart. When inserted on a seam or into a crack of a nut, they help split the shell apart.

The molars are much like ours, and squirrels use them to grind food. In fact, they grind food so much that it is very difficult to tell what they've been eating by examining stomach content or waste products. Watch how long a squirrel chews a single peanut.

There is a gap between the incisors and the back molars of the upper and lower jaws. It is called the **diastema**. If you find the skull of an animal and it has a diastema, you know the animal was a **herbivore**. Meat-eaters have canine teeth for tearing meat.

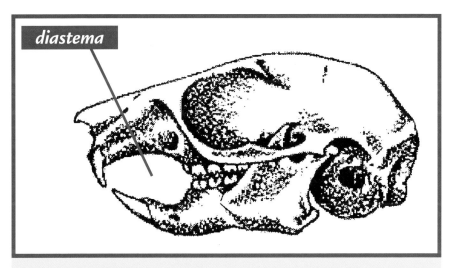

diastema

In this diagram of a squirrel's skull, you can see the diastema, which is a space between the teeth within the jaws.

The Squirrel's Anatomy

The squirrel's body and body functions evolved to fit life in trees, and so did the squirrel's entire lifestyle. How the squirrel reproduces, how and what it eats, how it finds shelter, and how it avoids predators have all evolved around trees.

Forests determined the squirrel's color. **Natural selection** in hardwood forests favored a gray color, which blends in well with the trees and ground cover. Squirrels in **coniferous** forests developed reddish or rust-colored fur. Today, both red and gray squirrels may live in the same habitat, especially in mixed forests, but in general, the pattern of gray in hardwood and red in coniferous forests holds true.

Squirrels' slightly bulging eyes, located high on a tapering face, give them a good sense of depth. This lets them race from branch to branch at high speeds to get away from a predator. Their eyes have yellow lenses that help cut down on glare, and the position of the eyes lets a squirrel detect the movement of a predator from almost any direction without moving its head.

A squirrel's keen sense of hearing protects it against surprise attacks, and its sense of smell is so exact that it can sniff out buried nuts and plant bulbs. Scent glands are located at the corners of the squirrel's mouth. These may be used to mark nuts before they are buried.

Although most squirrels are gray or red, in some places black squirrels and even white squirrels can be found.

WHITE SQUIRRELS

White squirrels are rare but they do appear occasionally. The cause is a gene **mutation** that inhibits the production of **melanin**—the substance that produces skin color, hair color, freckles, and suntans in humans. Without melanin, squirrels are born white.

Natural selection favors gray fur over white. So why do three towns—Kenton, Tennessee; Marionville, Missouri; and Olney, Illinois—have hundreds of white squirrels, far more than one would expect? People like the idea of having white squirrels around and encourage their reproduction and survival by providing nests and feeding stations for them. Because of this outside help, over time, the survival factor shifts in favor of the white squirrel. The town of Olney has a white squirrel logo on its official documents and on the arm patches of police and firefighter uniforms. It is illegal in Olney to run over a white squirrel. If you do hit one, you have to report it.

Four pairs of whiskers, deep in a squirrel's skin, serve as touch **receptors**. One pair is above the eye and another is under it. A third is around the nose, and the fourth is on the cheek. These whiskers are sensitive enough to let the squirrel feel even a slight breeze, a change of wind direction, or vibrations in a branch.

It is easy to see the similarity of squirrels' front feet to the hands of primates, including humans. Both squirrels and primates adapted to life in trees, which accounts for this great similarity. Squirrels and humans use the same muscles to grip a limb or a steering wheel. However, the human hand reflects a different course in evolution that resulted in humans' unique opposable thumb.

The toes of a squirrel's front feet are long and flexible. They are good for gripping thin twigs and picking up and handling nuts. All of the squirrel's long toes have needle-like, inwardly curved claws to grip

The hands of a squirrel look a lot like the hands of a human. The similarities are especially easy to see when a squirrel is holding a piece of food.

the bark of trees. The pads of the feet also help in gripping, and fur between the pads helps keep the feet warm. A very short **vestigial** thumb with a flat nail can be found on the front feet. Sweat glands on the feet release chemicals unique to each individual squirrel. This helps squirrels make their presence known to other squirrels.

The squirrel's tail is used for shade, warmth, communication, balance, as an umbrella, and even as a makeshift parachute. The hairs can be made to stand up, fluffing up the tail and increasing its warmth value. The word *squirrel* has its origin in Greek words that mean "shade tail."

If you watch a squirrel as it moves down a tree, you will notice that its back legs are rotated 180 degrees so that the back feet face backward and are pressed against the bark. This rotation is similar to what humans can do when we turn over our lower arm. The bones and muscles of the squirrel's leg changed over a long period of time to work as they do today. Other animals do not have this adaptation. Even though cats are good at climbing up trees, they often have to come down backwards.

The squirrel's skeleton also reflects adaptations to life in trees. The length of the thigh and lower leg bones allows for maximum support and, together with strong muscles, enables a squirrel to jump 4 feet (1.2 m) straight up from one branch to another and

When a squirrel is moving down a tree, its back legs rotate so that the back feet face backward, as seen here.

8 feet (2.4 m) from one tree to another. During a leap, the squirrel's front legs extend beyond its head. The legs are built to absorb the shock of landing on a hard tree trunk. Squirrels can leap even longer distances of up to 15 feet (4.6 m) when they leap from one tree into the lower branches of another. During these long leaps, the squirrel flattens its body and uses its tail as both a parachute and a rudder.

Many people have reported seeing squirrels take bad falls from great distances, appear stunned for a minute or two, and then recover. Squirrels' bodies are light, their muscles and bones are adapted to take shock, and they can increase air resistance to slow their fall a bit by flattening their bodies and fluffing out their tails. They often grab onto thin branches to break their fall. Even with these physical traits, falls can be deadly, and squirrels do suffer serious injuries if they land the wrong way or fall onto concrete. Falls occur most often when squirrels chase one another during mating season, when they are threatened by a predator, or when they are simply playing.

The average life span of a squirrel is 7 to 8 years. Some squirrels live twice as long in captivity.

Chase and Escape

Two members of the weasel family, the pine marten and the fisher, are a squirrel's worst nightmare. Once a pine marten starts to chase a squirrel, the odds are in the pine marten's favor even though the squirrel has the advantage of a better knowledge of the routes through the trees. The squirrel will risk jumps onto thin branches that might stop the marten, or it might rush into a hollow tree nest that is too narrow for the marten to enter. If the pine marten's intended victim is a flying squirrel, the chances shift in favor of the squirrel.

Flying squirrels have a curtain of skin connecting the front and back legs. These **nocturnal** animals

HOW FLYING SQUIRRELS "FLY"

Flying squirrels don't really fly, they glide. Like all squirrels, a flying squirrel takes time to study a jump carefully, moving its head up and down and sideways, and sometimes moving to a new location to check the distance it has to cover. It may leap from a crouching position on a horizontal branch or, if it is on a tree trunk, it may start the leap with its head facing downward.

It leaps and spreads its legs, catching air in its membrane. In flight, it can turn right or left by controlling the lift of air under its membrane. If it gets enough momentum, it can even glide upward for a short distance.

To land, it raises its front legs. The membrane catches air and brings the back legs forward. The squirrel raises its tail to slow down. All four feet come close to hitting the trunk at the same time, which softens the landing.

can leap 150 feet (46 m) or more. They are the only squirrels in North America that have a good chance of escaping a pine marten.

Squirrels of the Grand Canyon— How New Species Develop

A striking example of the effects of isolation on natural selection can be seen in two **subspecies**, or kinds, of squirrels that live on opposite sides of the Grand Canyon in Arizona. Both subspecies were originally members of a single species.

The Albert squirrel lives on the southern rim of the Grand Canyon.

Research shows that two populations of the same species were separated from one another. It first happened when the Grand Canyon formed and by changes in climate brought about by the last Ice Age, which ended about 10,000 years ago.

The smaller population of squirrels lived in an area of only about 800 square miles (2,072 square km) in what is now called the Kaibab plateau on the northern rim of the Grand Canyon. Here the strikingly beautiful dark Kaibab squirrel, with its dark underparts and white tail, evolved. On the southern rim, the Albert squirrel evolved. It has lighter underparts and a gray tail and occupies a broader area.

Both varieties have similarly long ears. In the 20,000 years or so since their separation, the two populations have not changed enough to be considered separate species. As more time passes and they remain apart, however, they may someday become separate species.

Chapter 2

Eating and Nesting

What Do Squirrels Eat?

THE EASTERN GRAY SQUIRREL'S first taste of spring is the new buds and shoots. You can see gray squirrels reaching out to the very tips of thin twigs high in the canopy of the forest. Later, they shift to fruits of maple, beech, elms, and nuts, including oaks, hickories, and pecans.

Squirrels also eat other, less well-known foods. They often eat fungus, especially when other foods are scarce. Fungus makes up more than 15 percent of the gray

Squirrels eat whatever food is available. They will even dig through garbage cans to find things to eat.

squirrel's diet in summer, and around 80 percent of the Albert squirrel's diet.

The squirrel diet varies with what is available. It includes a wide range of plant materials, such as apples, berries, rose hips, tulip bulbs dug from gardens, and the inner bark of trees. Squirrels may eat beetles, grubs, bird eggs, and sometimes even small birds. In hard times, they have been known to eat dead animals, including squirrels. Scientists don't believe that they actively seek out bird nests to rob them, but the squirrels happen to come upon them in their daily activities. The next time you see or hear a squirrel being mobbed by birds in a bush or tree, you can assume the attack is for good reason.

Bark stripping by squirrels can be a real problem in winter when there is a lack of food. Bark stripping may also be a way for squirrels to work off frustration caused by overpopulation or a lack of nesting sites. They also may like the sweet sap they can eat after the bark is stripped off.

Diets of Urban and Suburban Squirrels

City and suburban squirrels depend on trees, lawns, gardens, and bird feeders for food, but they also eat a wide variety of human food—much of which is harmful to their health. Squirrels have even been known to break into candy machines for sweets. Pepperoni pizza, hamburger buns, salty potato chips, and

Sometimes squirrels will strip the bark from trees. They may do it because they are frustrated or they may just like the taste of the sap underneath the bark.

other junk food reduce the squirrels' chances of survival. Many squirrels that live in city parks depend on humans to feed them, especially in the winter—and it shows. Some city squirrels prefer junk food to natural food. Compared to suburban or rural squirrels, city squirrels often look "ratty." The best snack we can give squirrels is unsalted nuts, and feeding them is a good way to get to know these interesting critters. Salted peanuts, however, can be harmful to squirrels.

In cities and suburbs, squirrels often enter bird feeders to get the birds' food.

Storing Food for Hard Times

A forest often produces a lot of food for brief periods of time. At such times, especially in the spring and fall, squirrels have more than enough food to meet their needs, and so they store food in a **cache** for the future. Storing high-energy nuts and building up fat reserves in their bodies enable squirrels to stay active throughout the winter and avoid having to **hibernate**.

A gray squirrel will usually go through a set procedure before it buries a nut. First it will sit and then use its "thumb" and "fingers" to rotate the nut around in its paws, licking it often. The sweat glands on the squirrel's paws and its saliva likely give the shell an odor that helps the squirrel find the nut again later after it has been buried. Experiments have shown that a squirrel can pick out a rotten nut, perhaps by judging its smell and its weight. If the nut is bad, the squirrel will put it aside.

To bury a nut, a squirrel digs a shallow pit about 1 inch (2.5 cm) deep, drops the nut in, and then shoves dirt and rocks or plant matter over the site with its nose and front feet. It will often gently pat the site before leaving. Young squirrels often do a poor job of burying, sometimes leaving the nut exposed. They get better at the skill after watching adults.

Squirrels never stop storing food: They keep on burying it for later even after they've gathered

Squirrels have a special technique for burying a nut. First they hold it and lick it in order to give it an odor they'll be able to smell later when they look for it again. Then they bury the nut in the ground.

enough to last through the entire winter. It's just part of their nature. In fact, "squirreling away" food for the future is so natural that a squirrel that has been reared in a cage will still go through the motions of burying food in the bottom of its cage.

Another Storing Technique

Gray squirrels prefer hardwood forests and bury single nuts, such as acorns, in a scattered pattern.

RIGHT-HANDED SQUIRRELS?

Some investigators claim that squirrels can be either right-handed or left-handed. A squirrel will hold a pinecone or other food by its ends. A right-handed squirrel will always hold the tip of the pinecone in its right hand. A left-handed squirrel will hold the tip in its left hand.

A squirrel usually buries its food in a shallow hole. Young squirrels have to practice a lot before they do a good job when they bury nuts for later.

Squirrels from the genus *Tamiasciurus*, such as the red squirrel and Douglas squirrel, prefer coniferous forests and store many cones in each cache. Unlike gray squirrels, squirrels of the genus *Tamiasciurus* are territorial and will defend their caches.

Red squirrels live in the pine, spruce, and fir forests and the mixed forests of much of Canada, Alaska, and the United States. In the summer, red squirrels can be seen high in pine trees nipping off green, unripened cones and letting them fall to the ground. When they come down from the tree, the squirrels carry the cones to hide them in hollow logs, between boulders, or in shallow holes formed by roots at the base of trees. The green cones ripen and serve as a

TERRITORIAL SQUIRRELS

Why do red squirrels store pinecones in caches and defend them, when gray squirrels bury individual nuts and don't defend them? One reason is that red squirrels put large amounts of food in one cache. They cannot afford to have it raided by other squirrels. In contrast, many mammals and birds eat acorns, and burying single acorns in a scattered pattern as gray squirrels do cuts down on loss to other squirrels and other animals. However, to defend territory containing a lot of scattered, single acorns would take up more energy than it would be worth.

Red squirrels are common in the pine, spruce, and fir forests of North America.

source of seeds for the squirrel to eat for as long as two years or more. These caches are relatively safe from other mammals and birds, which find cones difficult to open. Red squirrels, like many other squirrels, also store mushrooms and other fungi by stuffing them into the forks of tree branches and under the bark. Poisonous mushrooms don't seem to make them sick.

Recovering Stored Food

The gray squirrel is not territorial and so it does not spend time guarding its hidden food. It does not put all of its buried nuts in one area. Instead, it uses only a small area around any given tree and spreads its locations of food stores over a large area. The gray squirrel may chase away an animal if it happens to be nearby, but the squirrel does not stand guard over its food.

Experts believe that squirrels do not remember the site of individual caches, but they do remember the general area. In the winter, you can see a gray squirrel search an area in its usual zigzag pattern, sticking its nose in the snow and even burrowing beneath deep snow. When it smells a cache, it digs down to recover it. In a few studies, the rate of recovery of buried nuts was about 80 percent, which is pretty good.

Squirrel Nests

Those big bunches of sticks and leaves you see high up in trees are squirrel nests, or **dreys**. Usually, they are found in the upper third of the tree, located out from the main branches, or close to the trunk in the notch of a thick branch. They are rarely built at the very top of trees because of the danger of hawk attacks. And they are rarely found in lone trees, except in cities where there are few trees around. Squirrels build nests within jumping distance of trees or telephones so they can make fast getaways from predators such as raccoons, pine martens, snakes, opossums, and weasels. The locations and structures of dreys vary with the seasons and the weather.

Summer dreys are usually rather skimpy, temporary airy platforms located far out on the tree branches that receive less direct sunlight. Squirrels will often take an afternoon nap in these dreys. They may share a drey with several squirrels, especially siblings that are out on their own for the first time. Squirrels will also stretch out on a shady limb on summer days to cool off.

Winter dreys are larger and are positioned to catch the warmth of the sun. The drey where a mother raises her young is also well constructed. Dens in trees are favored in winter and for raising young. Squirrels will often enlarge the openings in trees made by woodpeckers into large birdhouses and convert them into dens. Several squirrels will often sleep together in the hollow of a tree in winter to preserve heat.

Drey Construction

Dreys are roughly spherical in shape, about 1 to 2 feet (30.5 to 61 cm) in diameter. First, a squirrel builds a platform of branches and twigs to serve as a support. The outermost wall of the drey is made of interwoven twigs, some with leaves attached, and accounts for

Squirrel dreys are generally tightly constructed shells designed to keep out bad weather and protect the squirrels from predators.

most of the size. An inner shell of tightly woven twigs and leaves, together with moss, grass, and bark, creates a strong watertight structure. Additional leaves are shoved into the spaces where they are needed.

Once the basic structure is built, an entrance into the center is made and a sleeping chamber inside is hollowed out by gnawing off and rearranging twigs and branches. The inner chamber is often lined with soft material, such as strips of bark, straw, moss, feathers, and fur. One report of a drey noted that a curtain of moss was hung over the inner entrance to the chamber.

The end result is a tightly constructed nest that is surprisingly windproof and waterproof. The squirrel continually repairs its nest, by plugging leaks and repairing any damage done by storms. A squirrel usually has several nests in case one is destroyed or becomes filled with fleas.

Chapter 3

Squirrel Family Life

Mating: The Chase

THE FEMALE SQUIRREL PICKS her mate by putting competing males though a difficult chase. In late winter or early spring, when a female is ready to mate, she gives off a distinct odor, which can be recognized by males up to a quarter of a mile (0.4 km) away. Males look for the female, and she allows them to enter her territory.

Next, one or two of the males begin to chase the female, and then the rest of the males join in. The sight of squirrels racing through the tops of trees, leaping from one branch to another, performing all kinds of aerobatics, and taking all kinds of risks—all with a lot of chatter—is a memorable scene.

As the female rushes through the trees, she leaves behind her scent so that the males will follow. At times, the female will turn and chase a male that gets too close. The chase may last a few minutes or an hour or more. Eventually, she gets tired and will mate. The lead male is usually the largest and strongest of the males, and this is the purpose of the chase ritual. The strong male can pass on his genes,

Before mating, the female squirrel leads the males on a wild chase. The strongest male, which ends up in the lead of the other males, is usually the one that gets to mate with the female.

which are most likely to aid in the survival of the off-spring. Mating usually takes place in the female's home territory. After a short time, she takes off again and the chase resumes. Over the course of a single day, the female may mate with several males.

The Single Mom

Female squirrels lead a tough life, especially eastern gray and fox squirrels, which usually have two litters a year, while the western gray, red, and Douglas squirrels usually have only one. The female raises her young by herself. After mating, the male has no role in raising the young.

Gray squirrels are pregnant for 44 to 46 days. They have their first litter in the early spring and then a second litter in the late summer. The number of babies in a litter varies from three to five. The babies are born naked, with their eyes shut, and weigh about half an ounce (14 g). They drink mother's milk for eight to nine weeks.

Young squirrels are covered by light fur by the third week, and their eyes open by the fifth week. During this time, nursing mothers must eat enough food to produce milk for their offspring. Many of the amazing acts squirrels perform in raiding bird feeders are made by female squirrels that are desperate for food. Just as male squirrels compete to mate, females compete for food.

Escaping From Danger

The female usually has more than one nest in case she has to move in a hurry or fleas take over her nest. She picks up her young around the belly and they curl around her mouth while they are carried. In emergencies, nursing young sometimes cling to their mother's nipples as she runs away.

To get as much food as they need, squirrels will take whatever food they can get.

Raising the Young

When baby squirrels are about seven weeks old, they take their first steps outside the nest. At first, they stay close to the nest. They move a bit farther away with each outing, and sample a variety of foods before returning to the nest to nurse. By around the ninth or tenth week of life, they are weaned from their mother's milk and really start to learn about their world.

Gray squirrels in northern climates may have two litters a year if food is plentiful. That means that the spring litter have only two or three months before they have to be able to fend for themselves as their mother prepares for her summer litter.

A squirrel has to gain a lot of knowledge and develop many skills if it is going to survive. Being able to hold onto tree bark is not the same as running quickly over thin branches to avoid a predator—even adult squirrels sometimes fall and injure themselves. Learning by watching and doing teaches the squirrel what it can and can't do.

During the time that her offspring are with her, the mother squirrel exposes them to a wide variety of foods. Squirrels are not born knowing how to open a nut. They have to learn. They also have to learn how to reach the acorns, maple seeds, and pinecones on the ends of thin twigs high up in trees.

Suburban squirrels have to learn to avoid traffic. They learn the safest routes through their neighbor-

Young squirrels learn from their mothers about different kinds of foods and sources of drinking water.

hood, using trees, fences, bushes, and roofs, just as humans get to know shortcuts where they live.

The mother squirrel helps teach her offspring who and what is a threat. Squirrels will often chatter most of the time. Plus, they will stomp their front feet at a predator, trespasser, or anyone who is interfering with their activity. They are often helped by other squirrels that join in making the predator's presence known to other prey.

Squirrels, especially yearlings, see cars as chasing predators. It is natural for them to reverse course at the last minute. Unfortunately, this does not work with cars, and it results in the deaths of many squirrels.

It's a tough world out there. Many young squirrels fall victim to predators, disease, accidents, cars, or starvation. One study showed a 50 percent death rate in squirrels that are born in the fall.

Youngsters on Their Own

About three weeks after birth, the bonds between offspring and their mother begin to weaken. Gray squirrels do not have strong feelings about territory. Although females with families may chase away other animals and aggressive behavior may be seen between competitors for acorns on the same branch, squirrels generally get along fairly well with each other. Unrelated gray squirrels may have nests in the same tree or in trees close by. Youngsters often sleep in another squirrel's nest as they move away from their family. In the winter, four or five squirrels may sleep together for warmth.

After a year of life, squirrels learn where and how to build a nest by watching other squirrels and by trial and error. Most quickly learn that sloppy nest construction results in a blown-down nest and a wet, troubling night. A nest built in a lone tree with no means of escape may also end in disaster. Most squirrels eventually get it right.

Leaving Home

By late September or early October, the litters born in the summer no longer need mother's milk, and the population of young squirrels is high. If an area will support more squirrels, some will stay; the others will move into a new area. If they are lucky, they have a healthy amount of stored fat in their bodies and a good supply of acorns and other nuts in their new neighborhood.

There are many difficulties in **migration** for the inexperienced, and natural selection takes its toll. Many squirrels are killed by cars when crossing highways, and many more will fall victim to predators, starvation, disease, and cold. The lucky ones will find a new place to live that has a healthy supply of food and will survive the winter by joining with other squirrels in dens or nests for warmth.

Sometimes, when excellent conditions produce many large and successful litters, there are mass migrations of hundreds and even thousands of squirrels. Mass movements have been recorded in Wisconsin and Ohio as thousands of squirrels have crossed roads and farms and swum across rivers and lakes. One observer recorded a squirrel climbing up an oar, entering a boat, and then walking out on the oar on the other side of the boat. Another person saw two populations of squirrels that were migrating in opposite directions meet in the middle of a river.

Chapter 4

Squirrels and People

Helping Squirrels

SQUIRREL NESTS ARE SOMETIMES destroyed by storms, predators, or even a male squirrel that wants to kill the young so that he might create a new litter. In these kinds of events, the young may fall to the ground. If you find a young squirrel, it's best to leave it alone, since the mother will usually rescue it. However, there are some things you can do to help.

If the squirrel is cold, you might warm it up by holding it close to your body or cupping in your hands. Then

If you find a baby squirrel, you can try to keep it warm by holding it in your hands.

return it to the ground. It is untrue, as many people believe, that the mother will be put off by your smell. Back off as far as possible and watch for the mother for an hour or so. If she does not return and you decide you are able to take care of the young squirrel, make a thorough search of the area, because other babies may be hiding in the leaves. Place the animals in a cloth-lined shoebox. Check on them often. It's best to get an adult to help you from the start.

Contact your state fish and game department, a local veterinarian, or search the Internet to find the nearest location of a Wildlife Rehabilitation Center that takes orphaned animals. A Wildlife Rehabilitation Center will advise you about giving temporary care, such as nursing. Some states issue a permit to adopt a squirrel, but always with the idea of releasing it.

Feeding Squirrels

Squirrels can be fun to feed. However, you should never touch them unless an adult is present and you know that the squirrel is healthy. If you want to watch a squirrel eat, first buy a good supply of nuts. Hazelnuts, walnuts, or pecans are ideal. Just make sure any nuts you use are unsalted. Corn kernels also work. Begin a pattern of leaving a few nuts in one place at a certain time. Don't leave too many. When you leave the food, make a "kissing" sound. The squirrels will begin to associate the sound with being fed. Eventually, the squirrels will be waiting for you each day. Stay a little longer each time until the squirrels come closer. Avoid any sudden movements. Don't bend down and offer a squirrel a nut with your fingers. Squirrels are nervous, jittery creatures and they might nip your finger in their rush to get the food.

Watching squirrels is a great way to learn about them.

It is fun is to get to know squirrels. Study their habits and find answers to any of your questions in books and on the Internet. In just a short time, you'll notice that each squirrel differs not only in shape but also in personality. This makes your studies much more interesting, since you will soon be able to tell one squirrel from another. You'll begin to notice all kinds of things, such as how they use their front feet to turn nuts and which

If you start feeding squirrels, they will get used to you and even demand food.

Corn is one of the squirrel's favorite foods.

foods each squirrel likes best. You'll notice that squirrels sometimes disappear for a few days and that if you crack a nut before giving it to them, the chances are better that they will eat it close by, rather than running off and burying it.

Once you begin a feeding pattern, squirrels will probably let you know when it's feeding time. They have been known to climb a screen door or sit on a fence and chatter loudly. Some people have reported that squirrels will even enter a house to remind their forgetful human feeder that it is time to eat.

Living with a Squirrel

Gregg Bassett and his wife, Kathy, of Elmhurst, Illinois, live with a squirrel named Happy that they got from a licensed breeder of wild animals. Happy sleeps and eats most of her meals in a large cage in a spare bedroom, but most of the time, she is free to walk around the house. She has learned to use the corner of her cage as a bathroom.

One of the first things Gregg and Kathy had to do was teach Happy to accept a variety of foods, such as sunflower seeds, corn, apples, bananas, grapes, cherries, cauliflower, and many other fruits and vegetables. Happy needed a lot of practice before she could easily open nuts.

Although some squirrels will chew on plants, Happy does not seem to have this bad habit. She does, however, like to bury nuts in a potted plant. In fact, she buries nuts all over the house—under pillows, in cushions, under the rug, behind curtains, in tool boxes, in apron pockets—you name it. If Gregg finds Happy's cache of food, Happy will chatter at him, just as she would do to another squirrel in the wild.

The loss of habitats is a real threat to the squirrel's survival.

Loss of Habitats

As cities get larger and larger, habitats are destroyed and populations of animals, such as squirrels, deer, rabbits, and groundhogs, are killed or become separated from one another by streets and buildings. Sometimes, the separate groups are united by corridors, such as railroad beds or streambeds, which allow the offspring to move to new places. In general, the loss of habitats is very dangerous to the survival of squirrels.

Food for Squirrels

Plants and Trees

Black oak
White oak
Turkey tail
Mockernut hickory
Black walnut
American beech
Black cherry

Flowering dogwood
Eastern redcedar
Eastern white pine
Virginia pine
Yellow poplar
Red maple
American holly

Insects and Animals

Red-backed salamander
Northern cardinal (nestlings)

Spring peeper
Luna moth

Trees Squirrels Use for Shelter

White oak
American beech
American elm
Red maple
Sweetgum
Black oak
American hornbeam
Mockernut hickory
Greenbrier
Silver maple

Spicebush
Smooth sumac
Pokeweed
Loblolly pine
Common elderberry
Switchgrass
Cinnamon fern
Willow oak
Japanese honeysuckle
Black locust

Predators That Hunt Squirrels

Great horned owl
Red-tailed hawk
Red fox
Black rat snake
Eastern hognose snake

Barred owl
Sharp-shinned hawk
Raccoon
Copperhead
Sarcoptic mange mite

HOW SQUIRRELS ARE CLASSIFIED
The Eastern Gray Squirrel

Kingdom	Animal
Phylum	Chordate
Class	Mammal
Order	Rodentia
Family	Sciuridae
Genus	Sciurus
Species	*Sciurus carolinensis**

*There are many different species of squirrels, but because the Eastern gray squirrel is one of the most common, it is used here as an example.

adapting—Changing over time to fit in better with the environment.

cache—A hidden supply of food.

carnivores—Meat-eating animals.

coniferous—Of or relating to evergreen trees or shrubs with needle-like or scale-shaped leaves.

dentine—Material that makes up the bulk of teeth. It is not as hard or dense as bone.

diastema—A space between the teeth in the jaw.

dreys—Nests or dens.

extinct—No longer living or living on Earth.

gnawing—Chewing with the teeth.

herbivore—Plant eater.

hibernate—Pass the winter in a resting state.

incisors—The cutting teeth in mammals.

melanin—Any of the various black, brown, red, or yellow pigments that provide color in animal skins.

migration—Movement to a new place, usually as the seasons change.

mutation—A change in the genes that leads to a change in a physical trait.

natural selection—A process by which those animals that are best adapted to their environment survive to reproduce.

nocturnal—Active at night.

organisms—Living things.

predators—Animals that hunt and eat other animals.

receptors—Cells or groups of cells that receive stimuli—signals from the outside world, such as sounds or touches.

rodents—Small animals that gnaw with their teeth.

subspecies—A group of animals within the same species that has unique characteristics.

vestigial—A body part that is small or not completely developed compared to the same body part in the ancestors of the same animal.

Bernett, R. J. "The Effects of Burial by Squirrels on Germination and Survival of Oak and Hickory Nuts." *The American Midland Naturalist* 98 (1977): 319–330.

Dagnall, J., J. Gurnell, and H. Pepper. "Bark Stripping Damage by Gray Squirrels in State Forests of the United Kingdom: A Review." *Ecology and Evolutionary Biology of Tree Squirrels*, ed. M. A Steele, J. F. Merritt and D. A. Zegers. Virginia Museum of Natural History, 1998.

Fitzwater, William D., Jr., and W. J. Frank. "Leaf Nests of the Gray Squirrel in Connecticut." *Journal of Mammology* 25(2) (1944).

Fox, J. F. "Adaptation of Gray Squirrel Behavior to Autumn Germination by White Oak Acorns." *Evolution* 36(4) (1982): 800–809.

Koprowski, John L. "Conflict Between the Sexes: Review of the Social and Mating Systems of the Tree Squirrels." *Ecology and Evolutionary Biology of Tree Squirrels*, ed. M. A. Steele, J. F. Merritt and D. A. Zegers. Virginia Museum of Natural History, 1998.

Long, Kim. *Squirrels: A Wildlife Handbook*. Boulder, CO: Johnson Books, 1995.

Muul, Illar, and John W. Ally. "Night Gliders of the Woodlands." *Natural History Magazine* 72(5) (1963): 18–25.

Smith, C. "The Evolution of Reproduction in Trees: Its Effect on Squirrel Ecology and Behavior." *Ecology and Evolutionary Biology of Tree Squirrels*, ed. M. A. Steele, J. F. Merritt and D. A. Zegers. Virginia Museum of Natural History, 1998.

Thorington, R. W., Jr., et al. "Aboreality in Tree Squirrels Ecology." *Ecology and Evolutionary Biology of Tree Squirrels*, ed. M. A Steele, J. F. Merritt and D. A. Zegers. Virginia Museum of Natural History, 1998, p. 119.

Jango-Cohen, Judith. *Flying Squirrels*.
    Minneapolis: Lerner Publications, 2004.

Spruch, Grace Marmor. *Squirrels at My Window: Life
    With a Remarkable Gang of Urban
    Squirrels*. Boulder, CO: Johnson Books, 2000.

Steele, Michael A., and John L. Koprowski. *North Ameri-
    can Tree Squirrels*. Washington, D.C.: Smithsonian
    Books, 2003.

Thorington, Richard W., Jr., and Katie E. Ferrell. *Squir-
    rels: The Animal Answer Guide*. Baltimore: Johns
    Hopkins University Press, 2006.

Web Sites

Squirrel-rehab.org
    http://squirrel-rehab.org/index.html

Squirrels.com
    www.squirrels.com

The Squirrel Lover's Club
    www.thesquirrelloversclub.com

The Squirrel Place
    www.squirrels.org/

The Wildlife Rehabilitation Information Directory
    www.tc.umn.edu/~devo0028/

ABOUT THE AUTHOR

James V. Bradley taught biology at Lake Forest High School in Lake Forest, Illinois, for 25 years. He also taught science in Colorado and in the United Kingdom. Bradley received the Illinois STAR Award (Science Teaching Achievement Recognition) in 1980 and was named by the National Association of Biology Teachers as outstanding biology teacher in Illinois in 1981. He retired from teaching in 1994, but continues to write and study science topics.